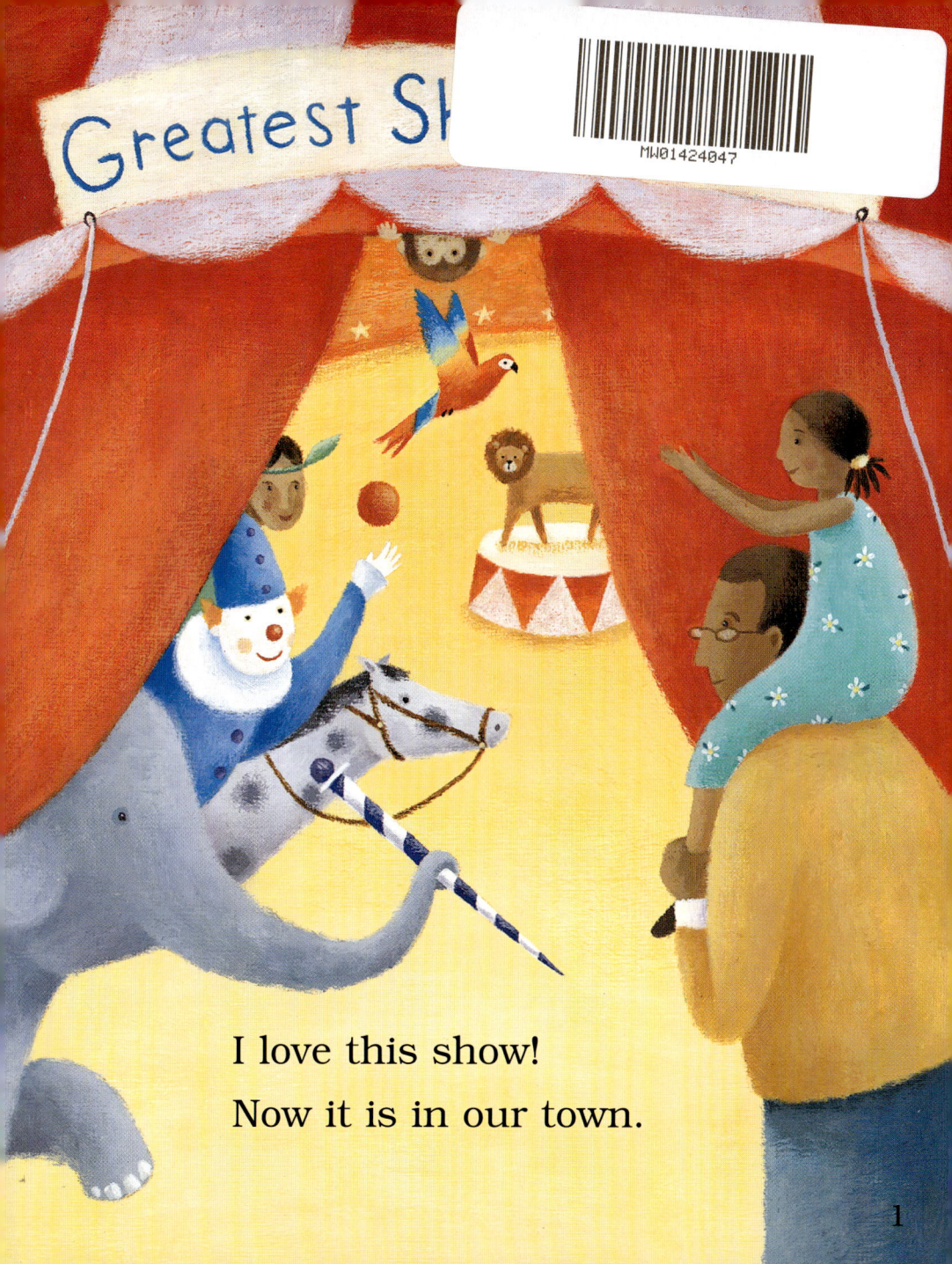

I love this show!
Now it is in our town.

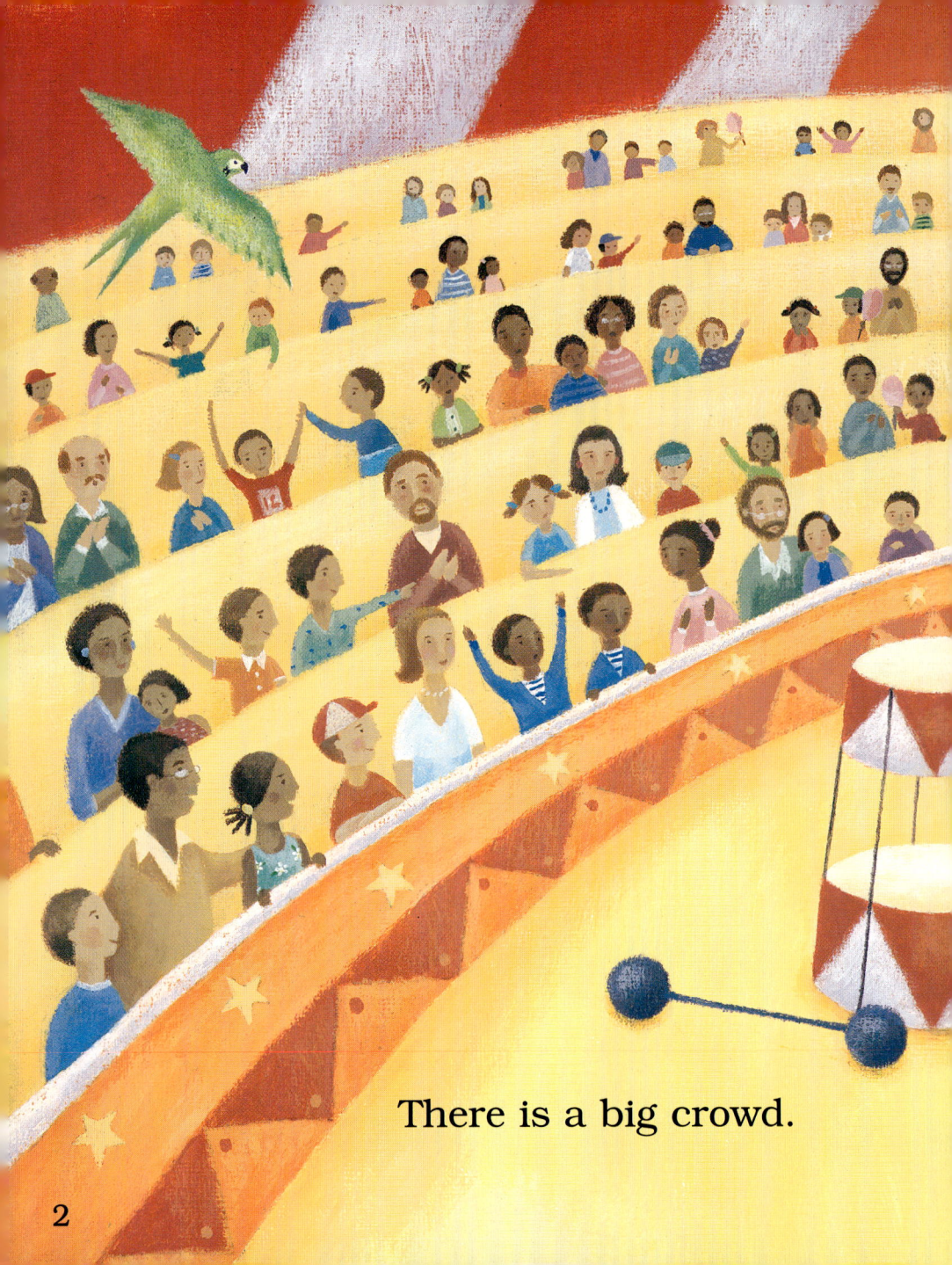
There is a big crowd.

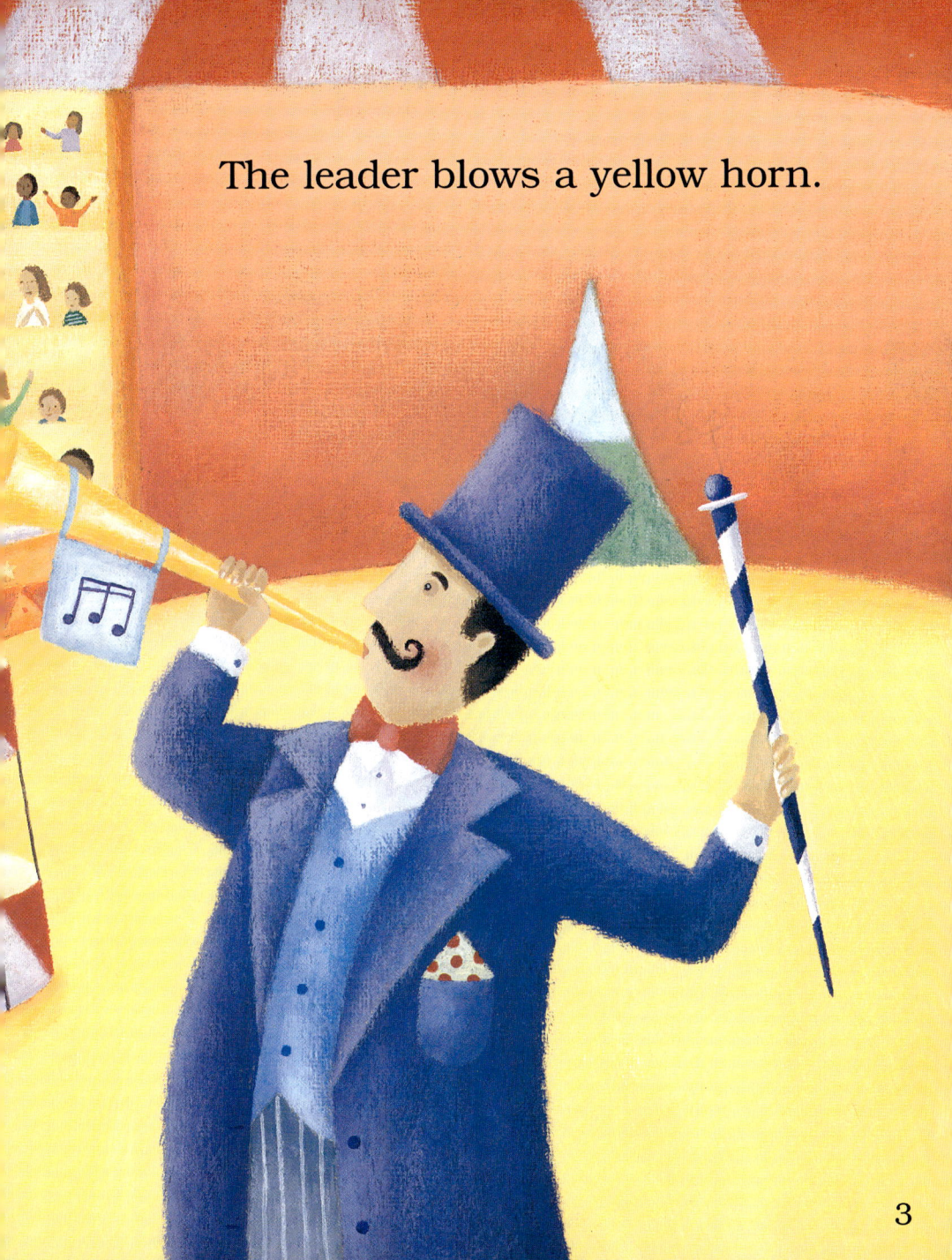

The leader blows a yellow horn.

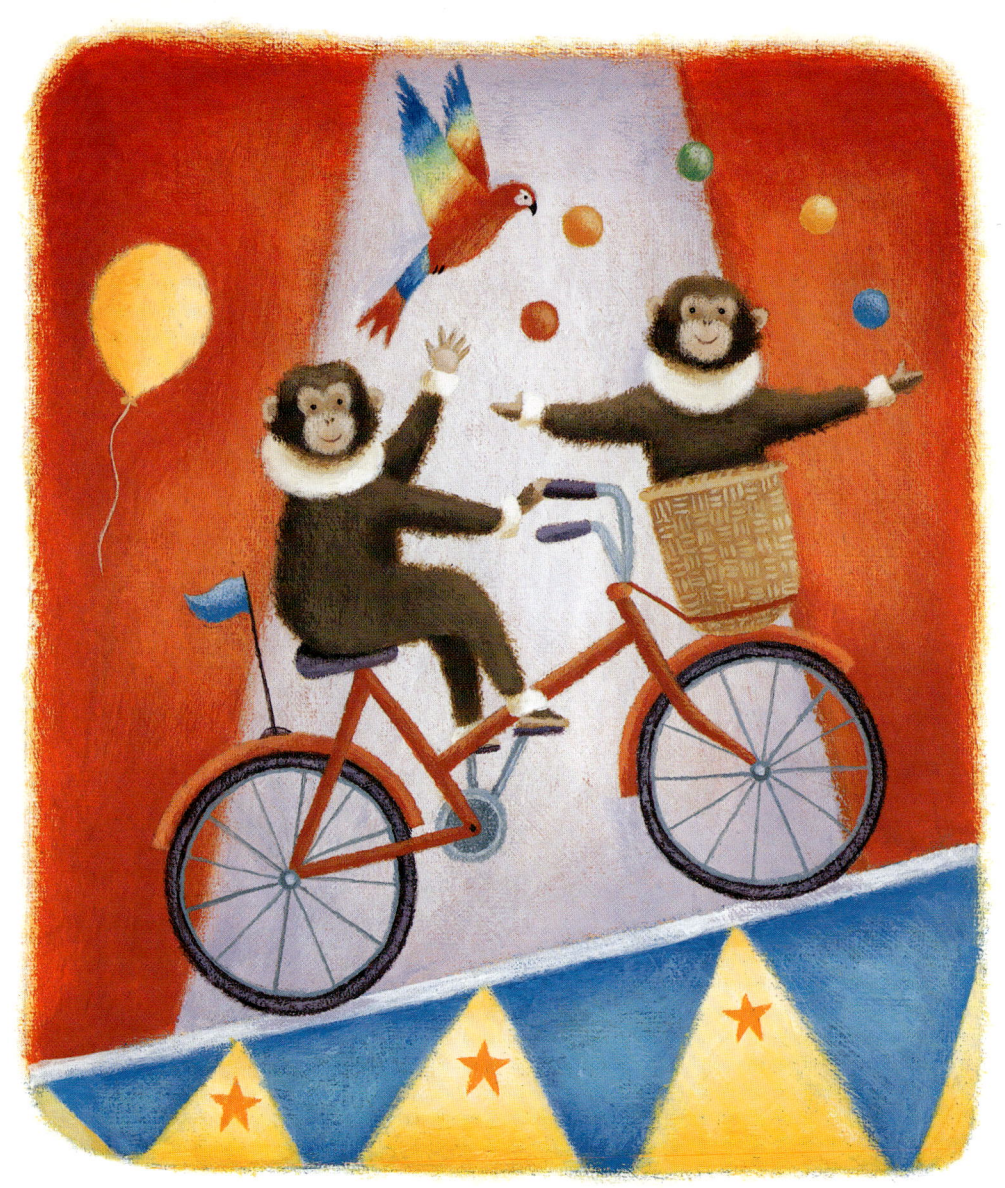

Two brown chimps do a tricky act.

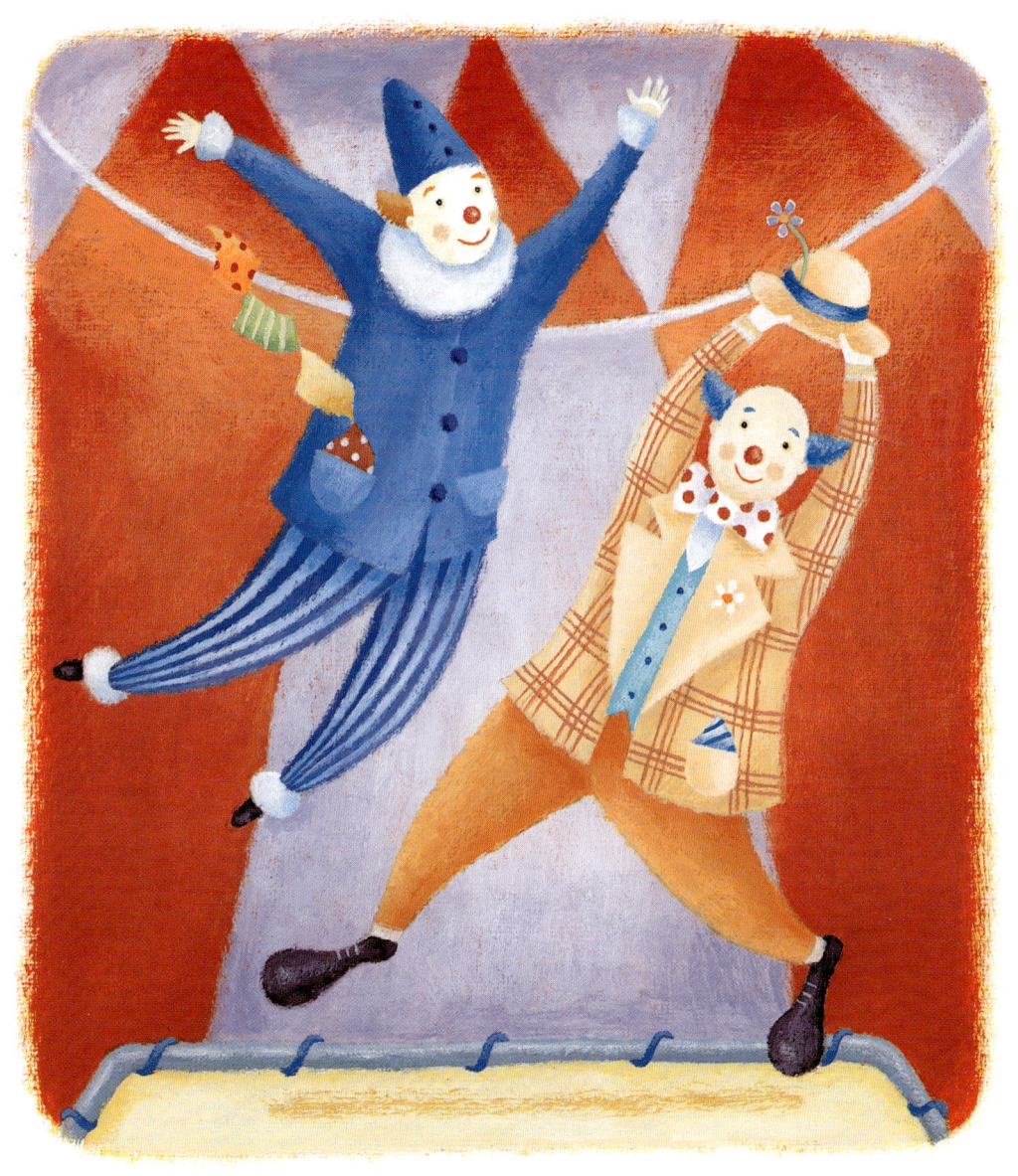

The clowns jump up and down.

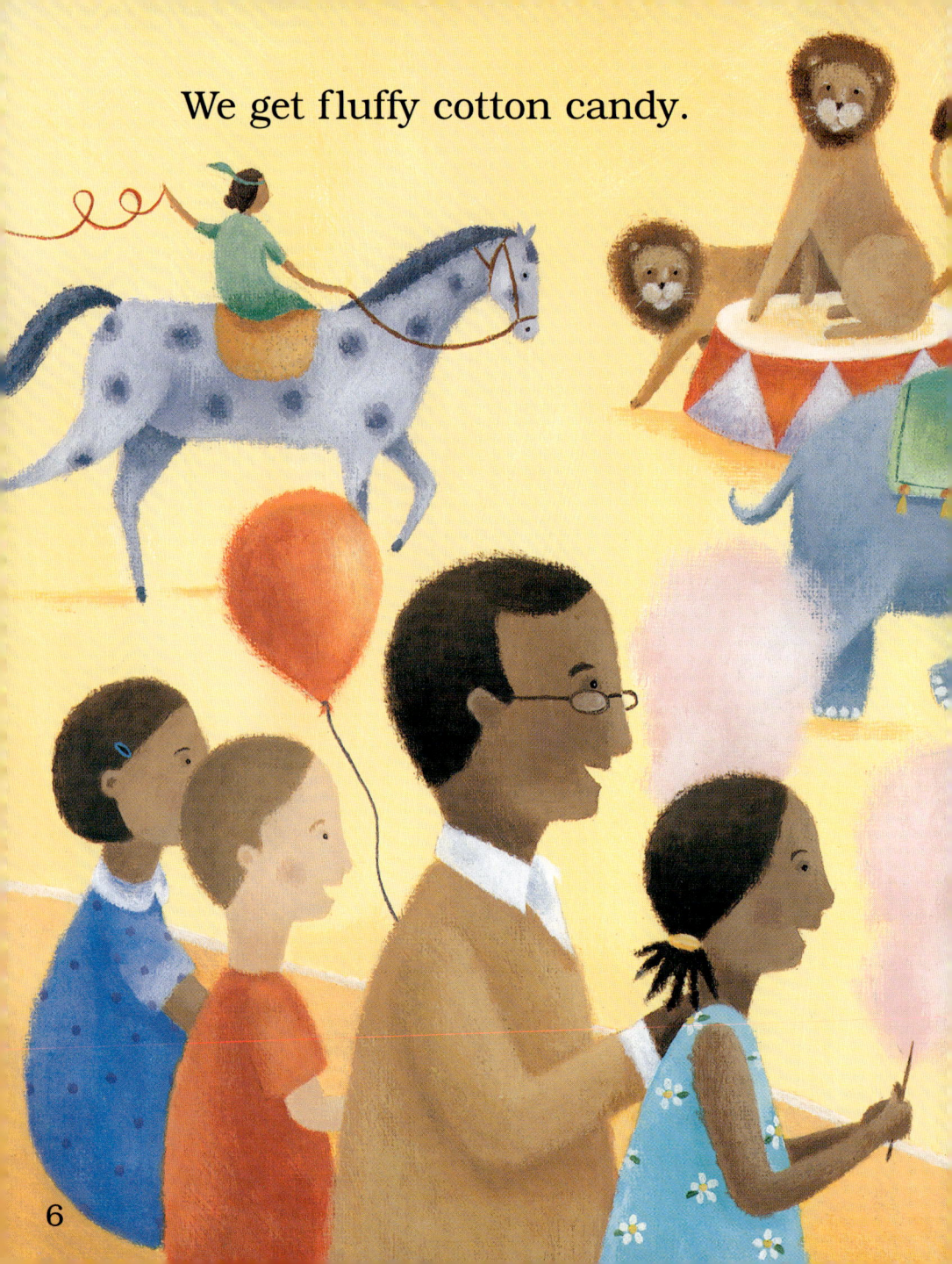
We get fluffy cotton candy.

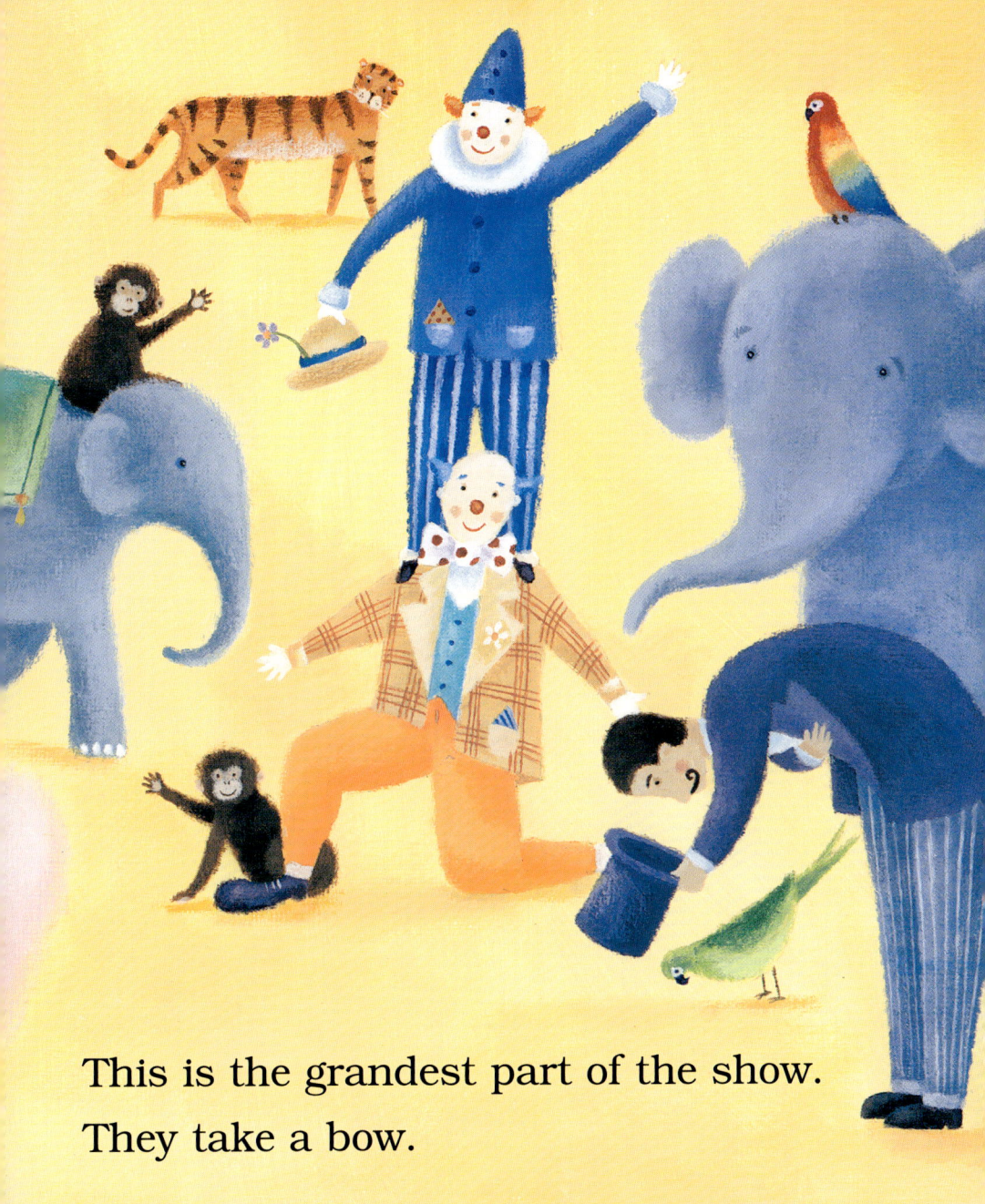

This is the grandest part of the show.
They take a bow.

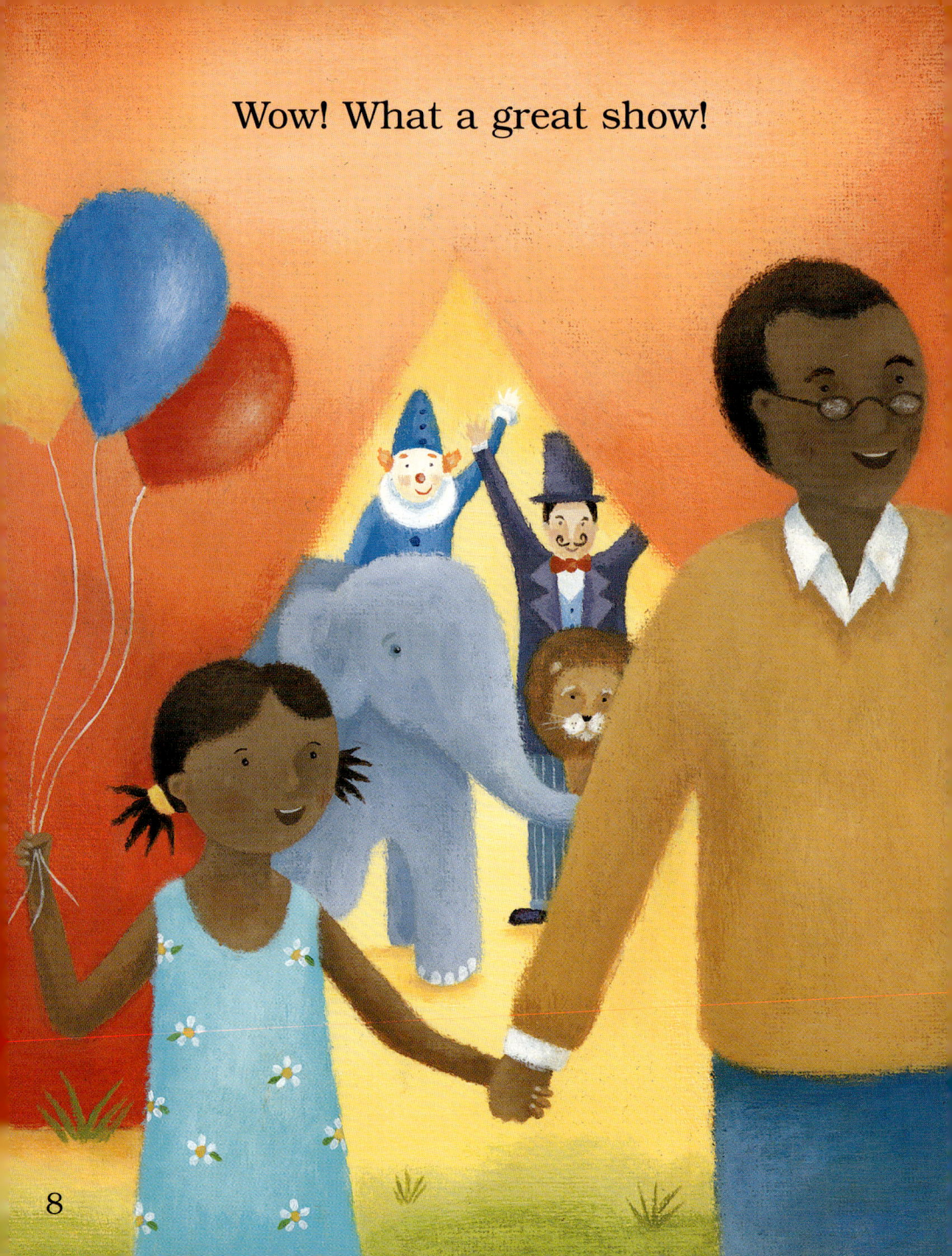